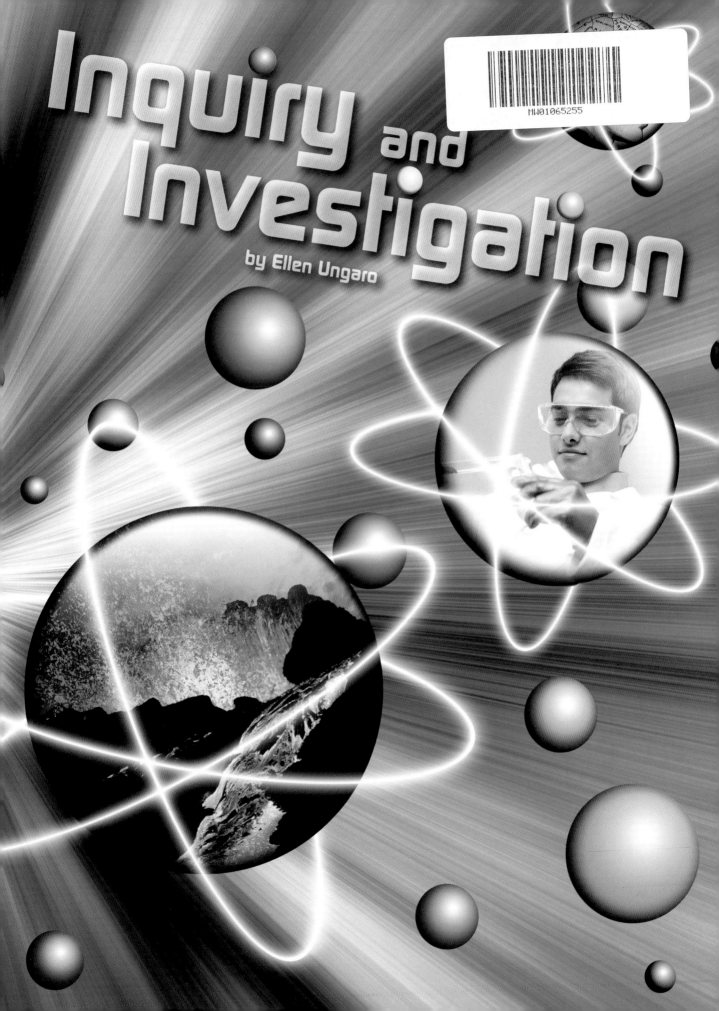

Table of Contents

Introduction
Science in Action 4
What do chimpanzees, particle colliders, and volcanoes have in common?

Chapter 1
What Is Science? 6
What are the goals and methods of science?

Chapter 2
How Do We Study Science? 16
What is the scientific method?

Cartoonist's Notebook 26

Chapter 3
Thinking Like a Scientist 28
What skills are important to the scientific process?

Chapter 4
Science and Society 36
How does science help people make decisions about the world?

Conclusion
How Scientists Work 42

How to Write a Lab Report 44
Glossary 46
Index 48

How is information about the natural world obtained and used?

INTRODUCTION

SCIENCE IN

Latest News ↓

Login

Password

Chimpanzees Create Sweeping Tool

Goualougo Triangle, Democratic Republic of the Congo—Scientists working on the Goualougo Triangle Ape Research Project have made an exciting new discovery! Using hidden cameras, they observed chimps uprooting plants. They then saw the chimps use their teeth to make the end of each plant into a brush-like tool. The chimps use the tool to sweep termites out of their nests. This sweeping tool is more complex than anything scientists have seen these animals use before. The chimps have created their own brooms!

The Big Bang Machine

Geneva, Switzerland—It took more than fifteen years, nine billion dollars, and 10,000 scientists and engineers from more than 100 countries to build it. Now that the Large Hadron Collider is ready, scientists hope this huge machine will provide clues about the formation of the universe. The collider was built outside Geneva by the European Organization for Nuclear Research. It is the world's largest particle collider. It lies in a circular 27-kilometer (16.8-mile) tunnel that is buried 50-175 meters (164-574 feet) below ground. Scientists will send beams of particles through ring-like tubes at nearly the speed of light until the particles collide. One of the goals is to mimic the conditions that existed just a fraction of a second after the big bang, when the universe began. Observing what happens when particles collide will help scientists understand the matter that makes up our universe.

ACTION

Volcanic Eruption!

Mount Redoubt, Alaska—At 10:38 P.M. on Sunday, Mount Redoubt in Alaska erupted, sending a plume of ash 9 miles (14.5 kilometers) into the sky. Over the next twenty-four hours, the volcano erupted five more times. Although Mount Redoubt had been quiet for twenty years, the Alaska Volcano Observatory began issuing eruption warnings several months ago. Scientists had noticed increased seismic activity around the volcano. The small earthquakes and tremors were a sign that magma was rising toward the surface. In an attempt to gather as much information as possible, scientists began taking frequent readings of the gases escaping from vents near the base of the mountain. They also made seven flights over the top of the volcano. Once the volcano has quieted down, scientists will collect rock and ash from the eruption. They will carefully analyze the samples to learn more about what caused the eruption and to help them make predictions in the future.

The scientists at work in all these stories are studying very different things—chimpanzees, particle physics, and volcanoes. But as you learn about scientific inquiry, you will see that these men and women all have the same goal: They want to understand the natural world.

Read this book to learn more about the goals and methods of science. Find out why science is important to us all. See if you have what it takes to ask scientific questions and find the answers.

Chapter 1

What is Science?

What are the goals and methods of science?

▲ Ancient people wanted to find explanations for mysterious events such as eclipses.

Essential Vocabulary

pseudoscience	page 13
scientific law	page 12
scientific theory	page 12
spontaneous generation	page 8

People have always had questions about the natural world. In ancient time, they wondered why the sun always rose in the east and why the seasons changed. Today, we still ask questions about the solar system. We ask questions about how to cure disease. We also ask questions about how our actions affect life on the planet.

Science is a way of understanding the world around us. Scientists ask questions about the world, and they look for answers. Scientific knowledge is the result of a great deal of debate and confirmation among scientists. New information can lead scientists to ask new questions. Sometimes, new information causes scientists to reject old ideas about how something works. The result is that scientific knowledge is always growing and changing. The theory, or idea, of spontaneous generation is a good example of how scientific knowledge evolves.

▲ Medical researchers around the world are trying to cure diseases such as cancer and AIDS.

CHAPTER 1

Early Observations Lead to a Theory

In ancient times, people observed that living things seemed to arise from nonliving things. For example, they observed that mice often appeared in moldy haystacks. They noticed that maggots, the worm-like creatures that grow into flies, often appeared on rotting meat. These observations led people to make certain conclusions. They believed that these living things arose spontaneously, or without cause. Mice grew spontaneously from haystacks. Maggots grew spontaneously from rotting meat. This theory was called **spontaneous generation**. This idea of living things growing from nonliving things was accepted for a long time.

Ancient Recipe for Bees

1. Kill a bull during the first thaw of winter.
2. Construct a shed and place branches and herbs on the floor.
3. Put the dead bull on the branches and herbs.
4. Wait for summer, when bees produced by the decaying body of the bull will appear.

▲ Long ago, people thought that bees could spontaneously generate from dead bulls.

In 1668, the Italian scientist Francesco Redi (fran-CHESS-koh RED-ee) (1626–1697) proposed a different explanation. He suggested that flies laid their eggs on the meat. Then the maggots hatched from the eggs. To test his idea, Redi conducted a simple experiment. He put pieces of meat into three containers. He left one container of meat open. He covered a second container with gauze. And he sealed the third container. As Redi predicted, no maggots appeared on the meat that was in the sealed container or in the gauze-covered container. The flies could not get to the meat. However, maggots swarmed all over the meat in the open container. Redi shared his findings with other scientists. They tried the experiment and got the same results. Some people began to accept the idea that life forms like maggots did not generate spontaneously.

▼ Today we know that flies do not generate spontaneously from rotting meat.

WHAT IS SCIENCE?

A New Tool and a New Theory

Around the same time Redi was doing his experiments, a Dutch fabric merchant was also hard at work. Anton van Leeuwenhoek (AN-tahn VAN LAY-ven-huk) (1632–1723) made his own magnifying lenses. He used the lenses to inspect the fibers in his fabrics. This led him to develop a new scientific tool, the microscope. Using this tool, people could see things invisible to the naked eye.

Looking through the microscope, Leeuwenhoek saw what he called "tiny animals." He was looking at microorganisms. These tiny life forms cannot be seen with the unaided eye.

When Leeuwenhoek shared his findings, other scientists began to study the "tiny animals." They discovered that some of the microorganisms caused meat, broth, and other foods to spoil. Where were these microorganisms coming from?

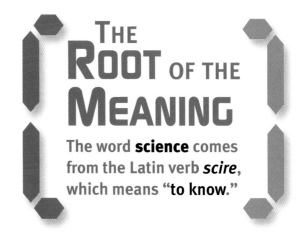

THE ROOT OF THE MEANING

The word **science** comes from the Latin verb *scire*, which means "to know."

The issue of spontaneous generation was fiercely debated for another century. Many scientists thought that these microorganisms spontaneously generated from food. But other scientists disagreed.

In 1859, the French Academy of Science had a contest. They offered a prize for the best experiment that either supported or refuted the theory of spontaneous generation.

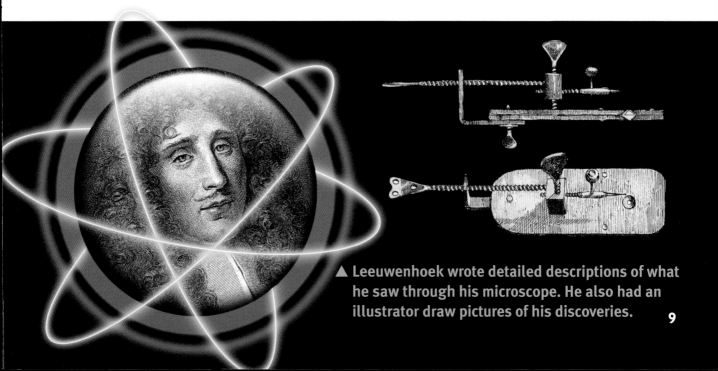

▲ Leeuwenhoek wrote detailed descriptions of what he saw through his microscope. He also had an illustrator draw pictures of his discoveries.

9

CHAPTER 1

Careers in Science: On the Hunt for Microbes

Scientists who study microorganisms do not spend all their time in the laboratory looking through microscopes. They also work in the field collecting samples. Recently, biologists from Rice University made an astonishing discovery in a cow field in Texas. They discovered the largest colony of amoebas ever—12 meters (39 feet) in length. Amoebas are single-celled microorganisms that usually live independently. It is rare to find them living together in a group. The discovery will help scientists learn how these microorganisms interact.

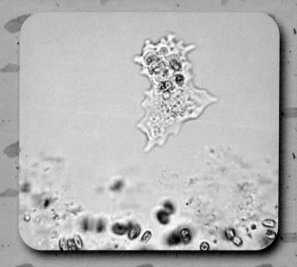

▲ Here is one amoeba magnified x800. Now imagine how many were in the colony!

The French scientist Louis Pasteur (loo-EE pas-TER) (1822–1895) set to work. He believed that the microorganisms that caused food to spoil did not spontaneously generate in the food. Rather, he thought that they were carried on dust in the air. But how could he test this idea? Pasteur designed an experiment to test his idea.

1. First, he put broth in a set of open flasks. He boiled the broth to kill any microorganisms already in it.

2. Then he put broth in a second set of flasks and boiled that broth, too. The flasks in the second set had a different shape than those in the first set. They had swan-shaped necks. This design meant that air could pass through, but any dust would be trapped in the necks.

The experiment supported his claims. The broth in the open flasks spoiled. The broth in the flasks with swan-shaped necks did not. Pasteur repeated his experiment many times to verify his results. Other scientists were also able to get the same results when they reproduced Pasteur's experiment. Pasteur's experiment did not support spontaneous generation. The experiment refuted the idea.

WHAT IS SCIENCE?

Everyday Science: Pasteurization

You can thank Louis Pasteur for the fresh milk you drink daily. He discovered that milk, juices, and other foods do not spoil as quickly if they are gently heated. This process of pasteurization also kills disease-causing organisms. Ultra-pasteurized foods are heated to a higher temperature and have longer shelf lives. Some foods are irradiated, or exposed to high-energy radiation. This method is called cold pasteurization.

CHAPTER 1

The Nature of Scientific Knowledge

It took more than 190 years and the work of many scientists to disprove the theory of spontaneous generation under Earth's present conditions. Each scientist built on the work of others. All of the scientists found evidence, or proof, to support their ideas. Other scientists tested and confirmed those ideas. When the evidence disproves an idea, scientists discard it—as they did with spontaneous generation. This continuous testing of ideas is how scientific knowledge grows.

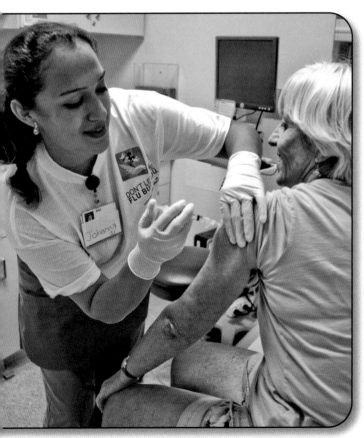

▲ Pasteur's germ theory led to the development of vaccines.

Scientific Theory

As you study science, you will learn that scientists have different ways to describe scientific knowledge. Scientists talk about scientific theories. In common language, a theory is an unformed idea. But in science, a theory is just the opposite. A **scientific theory** is a body of well-tested observations. A scientific theory helps to explain the natural world. It also helps us predict new findings. One example is Louis Pasteur's germ theory. It states that germs cause disease. Because Pasteur's theory explains what causes some diseases, it can be used to make predictions. We can predict that someone exposed to the flu virus will probably get the flu.

CHECKPOINT
TALK IT OVER

With a classmate, discuss how a scientific theory and a scientific law are alike and different, and why each is important to science.

Scientific Law

Scientists also talk about scientific laws. We have all types of laws in our society. Some laws are stated rules that tell us how to live in society. A **scientific law** is a different type of stated rule.

WHAT IS SCIENCE?

A scientific law explains something that has been shown to be true through countless observations. Newton's Laws of Motion are scientific laws. Isaac Newton developed the laws to describe the way objects move. One of Newton's laws states that for every action, there is an equal and opposite reaction. The liftoff of a rocket is an example of this law. As fuel is burned in the engine of the rocket, the exhaust gases escape downward and propel the rocket upward.

Scientific laws and scientific theories are not the same thing. A scientific law explains *what happens*. A scientific theory explains *why something happens*. A scientific theory does not become a scientific law.

PSEUDOSCIENCE

Do you consult your daily horoscope? Do you believe that the stars have an effect on your life or that they predict the future? If so, you are practicing astrology. Astrology is **pseudoscience**. In pseudoscience, things that might at first appear to be scientific do not meet the same standards as true science. *Pseudo* means false.

How can you identify pseudoscience? Unlike true science, there is usually no evidence to back up its claims, and the claims cannot be tested or confirmed by other scientists. Knowledge about a pseudoscience topic never grows or develops.

NEWTON'S LAWS IN ACTION

Newton's Laws of Motion explain many everyday occurrences. They explain why it is hard for a car to stop on an icy road and why planes can fly. Here's a quick way to see one of Newton's laws in action. Slide two coins together on a table. What happens when the coins collide? This is the law of action and reaction at work.

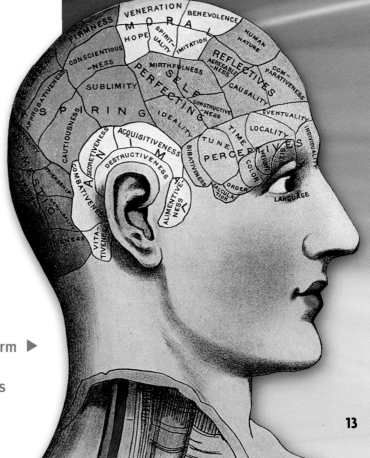

▶ Phrenology is the study of the size and form of a person's head in order to determine personality. It is a pseudoscience that was popular in the 1800s.

CHAPTER 1

Branches of Science

As the body of scientific knowledge has grown, so has the number of different areas of scientific study. To organize these different areas, science is often divided into three branches: life science, earth science, and physical science. The chart below shows the three branches of science and some of the different fields that scientists can specialize in. The photos show some of the work scientists in each branch do.

▼ A research chemist might work for a company that makes cleaners, food preservatives, or perfumes.

▼ Engineers work for car companies to design safer cars.

▲ Some botanists study plant diseases. Their work is important to farmers.

Branches of Science / Fields of Study

Branch	Definition	Field	Definition
Life science:	the study of living things, including plants, animals, and microorganisms	Botany:	the study of plants
		Ecology:	the study of living things and their environment
		Genetics:	the study of heredity
Earth science:	the study of Earth's properties as well as the development of Earth; space science is included in this branch of science	Geology:	the study of Earth, especially rocks, soil, and minerals
		Oceanography:	the study of the oceans
		Meteorology:	the study of weather
		Astronomy:	the study of the universe
Physical science:	the study of matter and energy	Physics:	the study of energy, forces, and motion
		Chemistry:	the study of the properties and interactions of matter

WHAT IS SCIENCE?

Summing UP

- Science is a way of understanding the world.
- Scientific knowledge is always growing and changing. Scientists build on the work of other scientists.
- A scientific theory is a body of tested observations that explains the natural world and can be used to make predictions.
- A scientific law explains something that has been shown to be true through observations.
- Today science is divided into three main branches: life science, earth science, and physical science. There are many fields in each of those branches.

Putting it All Together

Choose one of the research activities below. Work independently, in pairs, or in a small group. Share your responses with the class.

1 Research other examples of spontaneous generation. Then write a paragraph explaining how Pasteur's experiment would disprove each example.

2 Work with a partner to identify an example of a scientific law and a scientific theory. Using those examples, create your own definitions of the terms *scientific theory* and *scientific law*. Then show how the examples you have selected illustrate your definitions.

3 Choose a scientific field of study from the chart on page 14 that interests you. Then make a poster that defines the goals and different jobs of this field of study.

15

Chapter 2

How Do We Study Science?

What is the scientific method?

Francesco Redi had an idea about why maggots appeared in rotting meat. Louis Pasteur thought he knew why broth spoiled. But before their ideas could be accepted as scientific knowledge, the scientists had to provide evidence, or proof, to support the ideas.

The **scientific method** is a way for scientists to test and confirm their ideas. It is a process for asking and answering questions about the natural world. All scientific investigations follow an orderly series of steps. The scientific method you will now read about consists of seven steps.

▼ All scientific investigations follow an orderly series of steps.

Essential Vocabulary

control group	page 21
data	page 21
dependent variable	page 21
experimental group	page 21
hypothesis	page 19
independent variable	page 21
qualitative data	page 22
quantitative data	page 21
scientific method	page 16
variables	page 20

CHAPTER 2

The Scientific Method

- Make observations.
- Ask questions and do research.
- Develop a hypothesis.
- Design and conduct an experiment.
- Collect and analyze data.
- Draw conclusions.
- Communicate results.

Make Observations

Science begins with an observation. An observation is information you gather by using your senses. You use your sense of sight, smell, touch, taste, and hearing. Some observations about the world can be made directly. For example, you can very easily observe that it is a warm night just by going outside. Other observations are made by using tools that extend the senses. For example, thermometers can tell us exactly how warm it is. Telescopes can help us explore exactly which stars are in the night sky.

Ask Questions and Do Research

Do you enjoy making and eating popcorn? If so, you may have noticed that popcorn smells and tastes good. You also may have noticed that some kernels never pop. You have made observations based on your senses of sight, smell, and taste. Observations often lead to questions. Perhaps now you may wonder why you like popcorn or why certain kernels never pop. Although all questions are good, not all questions are scientific.

The distinction is that a scientific question can be tested. For example, "Why do I enjoy eating popcorn?" is not a scientific question. You cannot really test all the reasons why you like eating something. On the other hand, "Why don't all of the kernels pop?" is a scientific question. You can test this question. A good scientific question is not necessarily the first question that occurs to you. You might need to do more observing. You also need to research the problem. A final question might be "What factor prevents some popcorn kernels from popping?"

Checkpoint
Talk It Over

Discuss the following questions with a partner. Decide which are scientific questions and which are not. Explain your reasoning.

- How long can plants live without sunlight?
- Which contains more caffeine: soda or coffee?
- Are cows smarter than pigs?

HOW DO WE STUDY SCIENCE?

Develop a Hypothesis

Once scientists have a question, they will want to plan an experiment or an investigation to answer the question. Before scientists do this, however, they first need to state a hypothesis. A **hypothesis** describes what scientists expect to discover. It is often an *if-then* or cause-and-effect statement. You can think of a hypothesis as an educated guess about the outcome. A hypothesis is not a random guess; it is based on observations, prior knowledge, and research.

Some hypotheses are tested by making more observations. Others are tested by doing experiments. In either case, a good hypothesis is proposed in such a way that it can be tested to determine if it is or is not supported by the evidence.

Consider your popcorn again. You want to determine the cause of unpopped kernels. You have observed that some kernels are smaller than others. You have also done some research and learned that popcorn needs a certain amount of moisture inside the kernel in order to pop. Which of these two factors—kernel size or kernel moisture—do you want to explore? You decide that you want to find out how kernel moisture affects popping. Your hypothesis should state what you expect to discover: *If the amount of moisture in popcorn kernels is related to popping ability, then overdried popcorn will not pop as well as regular popcorn.*

Meet the Scientist: Backyard Astronomer

Powerful telescopes like the Hubble Telescope are helping astronomers see farther into space than ever before. But Robert Evans has made some amazing discoveries using just a backyard telescope and careful observation. Evans, a minister from Australia, holds the world's record for spotting supernovae. These exploding stars appear as bursts of light in distant galaxies and then disappear within a few months. Evans, who has been studying the night sky since he was a teenager, spotted his first supernova in 1981. Since then he has found forty-one more.

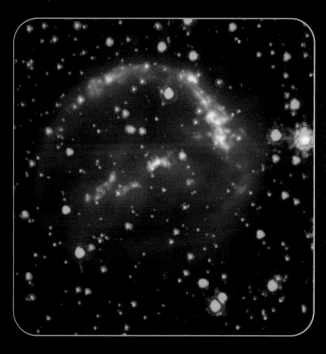

▲ To find a supernova, sky watchers need to be very careful and patient.

CHAPTER 2

EVERYDAY SCIENCE

Corn plays an essential role in everyday life. Corn syrup sweetens countless beverages and foods. Cornstarch thickens soups and many other prepared foods. Cornstarch is also used in wallpaper glue, cardboard boxes, and plasterboard for building. In the United States, corn is a major source of ethanol, which is added to gasoline to reduce air pollution.

▲ Corn is the most widely grown crop in America. Corn plants are a type of grass.

Design and Conduct an Experiment

Once scientists have formulated a hypothesis, they can plan an experiment to test it. And so can you! An experiment is a means of gathering information under controlled conditions. To test your hypothesis, you can dry popcorn by letting it sit out for a week. You can also put popcorn in an oven for two hours at a low heat of 93°C (200°F) to reduce the level of moisture in the kernels. Then you can pop the overdried popcorn and regular popcorn and observe the results. But to design a proper scientific experiment, you need to consider a few more things.

All experiments have **variables**, or factors that can be changed. In an experiment, a scientist wants to test only one variable at a time, keeping all other factors constant, or the same.

When you make popcorn, many factors can affect the number of unpopped kernels that are left. You can make popcorn in a microwave, in a hot-air popper, or on the stove using oil. Making a big batch of popcorn versus a small batch of popcorn might lead to a different result. These factors—what you pop the corn in, and how much you make—are variables.

To design your popcorn experiment, you need to eliminate these variables. So the experiment should use the same brand of popcorn, the same amount of popcorn, and the same popping method for each trial. You want to be able to test just the dryness of popcorn.

HOW DO WE STUDY SCIENCE?

In this experiment, the popcorn's moisture content is the independent variable. The **independent variable** is what the scientist is controlling. The number of unpopped kernels at the end of the experiment is the **dependent variable**. In other words, the number of unpopped kernels *depends* on the moisture content of the popcorn.

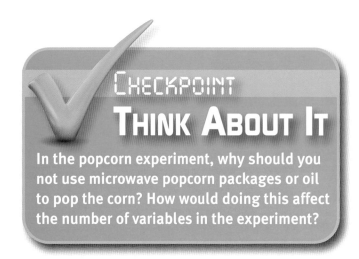

Checkpoint
THINK ABOUT IT

In the popcorn experiment, why should you not use microwave popcorn packages or oil to pop the corn? How would doing this affect the number of variables in the experiment?

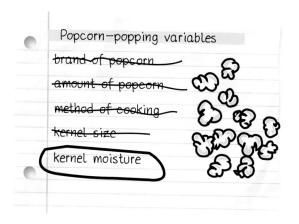

An experiment also needs a **control group**, where all variables are kept constant. In this experiment, you will use regular popcorn as the control group. The dried popcorn is the **experimental group**. After popping both the dried popcorn and the regular popcorn, you can compare the results from the control group and the experimental group.

You also need to pop several batches of both the control group and the experimental group. A scientist wants to repeat a test to make sure that the results are always the same.

Collect and Analyze Data

During and after an experiment, scientists make observations in order to collect and analyze data. **Data** are various types of information. Experiments and investigations usually provide two kinds of data.

Quantitative data are numbers and measurements. For example, a meteorologist studying the formation of hurricanes would gather measurements such as air temperature, water temperature, and wind speed.

Comparison of Unpopped Kernels in Dried and Regular Popcorn			
DRIED POPCORN		**REGULAR POPCORN**	
Batch	# of unpopped kernels	Batch	# of unpopped kernels
A1	15	B1	5
A2	12	B2	8
A3	18	B3	3
Average	15	Average	5.3

▲ The quantitative data collected from the popcorn experiment might look like this.

CHAPTER 2

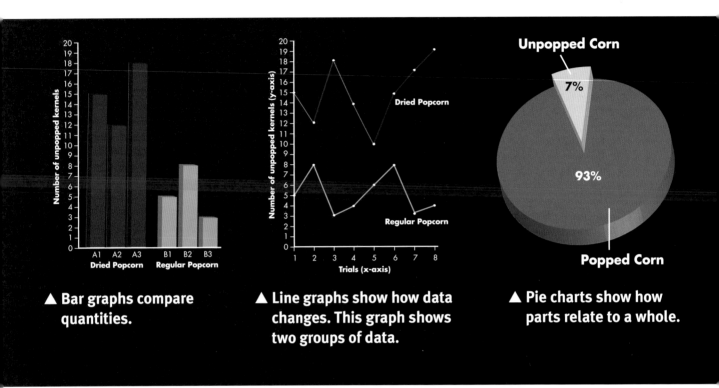

▲ Bar graphs compare quantities.

▲ Line graphs show how data changes. This graph shows two groups of data.

▲ Pie charts show how parts relate to a whole.

Qualitative data are generally descriptions. A biologist working in the field studying the behavior of gorillas would collect data describing where the gorillas find food, when they eat, and how they interact with one another.

Scientists then analyze the data that they collect. They can read the data from data tables, or they can represent the data visually as bar graphs, line graphs, or pie charts. Each type of graph emphasizes different features of data.

Draw Conclusions

At the end of an experiment, scientists will determine if the experimental results *support* or *do not support their hypothesis*. Scientists will not say that an experiment proved or disproved the hypothesis. This is because science is never totally certain. No one can test all possibilities. New experiments and data may or may not support the hypothesis.

When the results of an investigation support a hypothesis, they can often lead to new questions. Suppose your popcorn experiment supports your hypothesis. It supports that dried corn will have more unpopped kernels. What new questions might you ask? What other tests could you do to better understand the "popability" of popcorn?

If experimental results do not support a hypothesis, then the scientists can use the data to develop a new hypothesis to test. Even experiments that do not support a hypothesis help to advance science. Good experiments eliminate failed hypotheses and create new questions.

Whatever the outcome, scientists will repeat their experiments. They want to make sure that the results are consistent over time.

Communicate Results

Science advances rapidly because scientists communicate their results. As they work, scientists keep detailed notes. Scientists share their work when they share these notes, or publish research papers in scientific journals and online. Scientists also share their work with others when they teach or speak at conferences. In this way, new knowledge is spread rapidly. Scientists everywhere can build on the work of others and ask new questions based on that work.

Communicating results is also important because it allows other scientists to repeat experiments and confirm the results. Science depends on evidence. A scientist's ideas have to be tested and confirmed by other scientists to be considered valid.

Try It: Analyze the Data and Draw Conclusions

Analyze the data collected from the popcorn experiment shown in the table on page 21. Do you think the experimental results support the hypothesis?

HOW DO WE STUDY SCIENCE?

The Scientific Method at Work

In the real world, scientists may not always follow every step of the scientific method in order. Not all scientific questions can be answered by an experiment with a control group and an experimental group. Scientists studying volcanoes or earthquakes, for example, cannot create a control group. Some investigations will require that scientists collect qualitative data over a long period of time. But no matter what they are studying, all scientists test their ideas and find evidence to support those ideas.

▲ Before the invention of the printing press, it took years to spread scientific theories and findings. Today, we can share new scientific knowledge instantly.

Hands-On Science: Nature's Oxygen Generator

Plants are important to the atmosphere because they release oxygen into the air as a product of photosynthesis. During photosynthesis, plants use light energy to combine water and carbon dioxide to form a type of sugar and oxygen. The amount of light energy a plant receives affects the rate of photosynthesis and the oxygen production.

Problem
How can the rate of photosynthesis under varying light conditions be measured in terms of oxygen production?

Hypothesis
If the rate of photosynthesis is related to the amount of light energy a plant receives, then a plant that gets more light should produce more oxygen.

Materials
- battery jar
- water
- *elodea* (pond weed)
- funnel
- large test tube
- lamp with 60-watt bulb
- glass-marking pencil
- metric ruler

Procedure
1. Fill the battery jar 3/4 full with water.
2. Place two or three sprigs of *elodea* in the jar and cover them with an inverted funnel. The stem of the funnel should be below the water level.
3. Fill a large test tube with water from the battery jar.
4. Place your thumb over the mouth of the test tube, turn it upside down, and lower it over the stem of the funnel. Keep the test tube opening below the level of water in the battery jar so the water in the test tube does not run out.
5. Place a lamp with a 60-watt bulb at a distance of 10 centimeters (4 inches) from the battery jar. After 10 minutes, carefully remove the test tube from the funnel's stem by placing your thumb over the mouth of the test tube and lifting it out of the jar.
6. Draw a line on the test tube between the collected gas at the top of the test tube and the water level below the gas. Measure the height of the column.
7. Repeat this procedure with the lamp placed at a distance of 20 centimeters (8 inches), then at 30 centimeters (12 inches), 40 centimeters (16 inches), and finally at 50 centimeters (20 inches).
8. Construct a table to show the collective data, then use this data to construct a line graph.

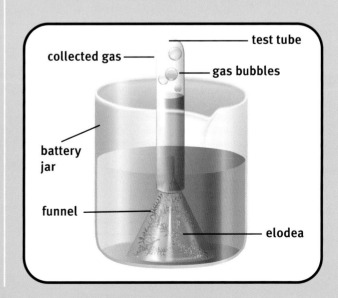

HOW DO WE STUDY SCIENCE?

Summing UP

- The scientific method is a process of asking and answering questions about the natural world.
- By making observations, asking testable questions, developing hypotheses, designing and conducting experiments, analyzing data, drawing conclusions, and communicating their results, scientists can make great advances in science.

Putting it All Together

Choose one of the research activities below. Work independently, in pairs, or in a small group. Share your responses with the class.

1 Work with a small group to record observations about a plant. Use your senses of sight, touch, smell, and hearing. From your observations, ask a scientific question about the plant. Then use the question to state a hypothesis that is testable.

2 Chapter 1 described the work of Francesco Redi, Anton van Leeuwenhoek, and Louis Pasteur. Choose one of the scientists. Research his work. Then write a paragraph describing how his work is an example of the scientific method. Identify his hypothesis and describe his experiment, observations, and conclusion.

3 Design a poster that illustrates a well-known experiment and also explains the steps of the scientific method.

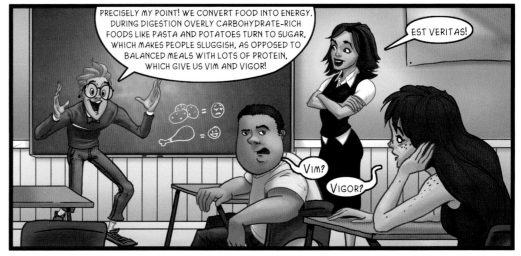

Science is about asking questions and finding possible answers.

Have you ever had a problem? Or asked a question about why something is the way it is?

What was your problem or question?

Can you think of a solution or an explanation?

Can you develop an experiment to test your reasoning?

Why or why not?

CHAPTER 3

THINKING LIKE A SCIENTIST

WHAT SKILLS ARE IMPORTANT TO THE SCIENTIFIC PROCESS?

People use science skills every day without realizing it. You use science skills when you take note of things around you. You use science skills when you ask questions and get answers. You use science skills when you solve problems. You also use science skills when you try to explain something to others. You do not need a scientific problem to practice science skills. You can sharpen your science skills simply by interacting with and learning about the world around you.

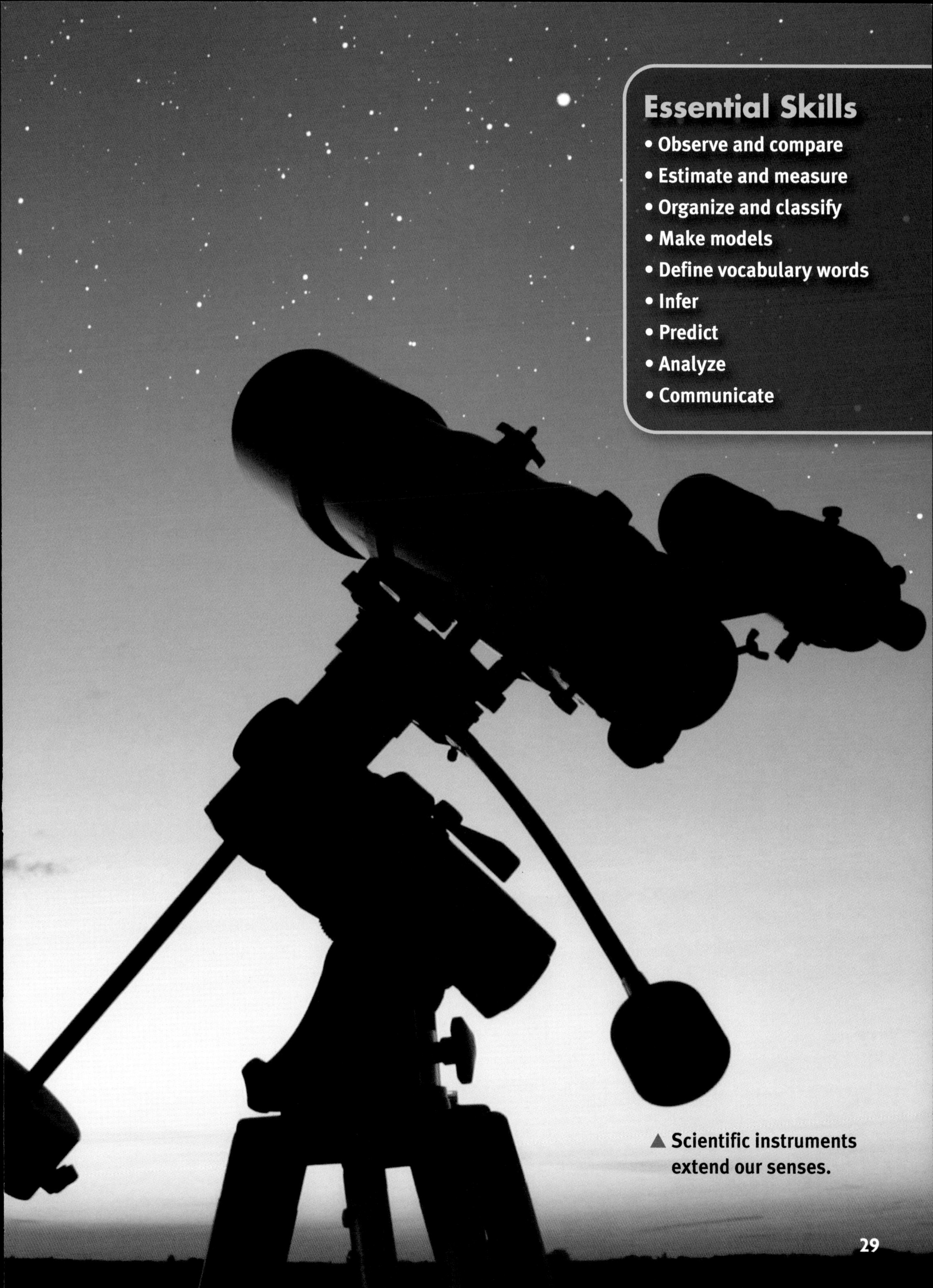

Essential Skills

- Observe and compare
- Estimate and measure
- Organize and classify
- Make models
- Define vocabulary words
- Infer
- Predict
- Analyze
- Communicate

▲ Scientific instruments extend our senses.

CHAPTER 3

Observe and Compare

Scientists often begin by observing—it might be an object or it might be an event. When you observe something, you can use all of your senses. If an unusual bird flew above the schoolyard one day, you could describe what it looked like. You could describe the kind of noises it made. You could watch the way it flew. You could also compare it with other birds you know. You might notice that it is bigger than an owl but not as big as the bald eagle you once saw. All of your observations can help you identify the bird.

Estimate and Measure

Estimating and measuring are skills you use every day. A simple task such as cooking dinner involves both of these skills. The cook needs to measure the ingredients in a recipe and set the temperature of the oven. The cook also needs to estimate how much food will be needed. If six people are eating dinner, the cook has to guess about how much lettuce will be needed in order to make enough salad for all.

▼ We can use observations to develop a hypothesis and to compare results.

Organize and Classify

Scientists gather information through observations and experiments. To make the information meaningful and useful, they need to organize and classify it. For example, biologists classify plants and animals by their similarities. In the animal kingdom, scientists divide animals into different groups based on their similar characteristics. Then scientists continue to group animals into smaller and smaller categories. Those animals sharing the smallest group are very much alike.

When you do research and write a report, you organize and classify information. If you are doing research about a particular country, you might organize the information according to categories such as history, geography, economy, climate, and interesting sites. You would put all of the information about the country's geography together and all of the information about the country's history together. This would help you write a clear, concise, and complete report.

THINKING LIKE A SCIENTIST

MEET THE SCIENTIST: THE BUG COLLECTOR

Rampa Rattanarithikul (RAHM-puh rah-tah-nuh-REETH-ih-kul), a scientist in Thailand, has spent more than fifty years studying mosquitoes. She has collected and carefully preserved specimens of these animals. She has also classified more than 400 different species. Her work has helped other scientists who study diseases that are spread by mosquitoes.

CLASSIFICATION OF MODERN HUMANS

Kingdom	Animalia	Large group of organisms comprising only animals
Phylum	Chordata	Animals that have a notochord (precursor to spinal column)
Class	Mammalia	Animals that feed milk to their young and have hair
Order	Primates	Animals with forward-facing eyes, opposable thumbs
Family	Hominidae	Animals that walk on two legs (and knuckle-walkers)
Genus	*Homo*	Humans
Species	*Homo sapiens*	Modern human beings

▲ We organize and classify animals using data gathered by observation and experiments.

CHAPTER 3

✓ Checkpoint
Think About It
How many words can you think of that use one of the roots in the chart on page 33? Work with a partner to make a list.

Hands-on Science: Make a Scale Model

Models can help you get a better sense of size and distance. You have probably seen diagrams that show Earth and its moon relatively close together and of a similar size. Creating a scale model of the two will provide a more realistic idea of their sizes and the distance between them. You'll need a white balloon for the moon and a blue balloon for Earth. Blow up the blue balloon so that it has a circumference of 60 centimeters (24 inches). Blow up the white balloon so that it has a circumference of 15 centimeters (6 inches). Place the Earth balloon in the center of the room. Place the moon balloon 6 meters (20 feet) away.

Make Models

A model is a visual representation of something. A model can be a conceptual representation (diagram, picture, computer image) or physical representation (three-dimensional). Models are useful because they can help you understand an object or a process. A globe is a model of Earth that enables you to see the entire planet and the relative sizes of its features. You can make models to help you learn about the solar system, a plant or animal cell, a human organ such as the heart, or a tornado.

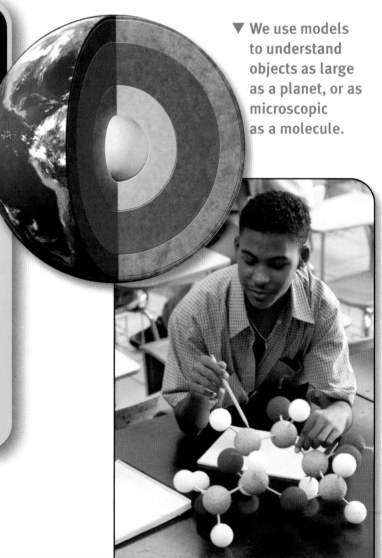

▼ We use models to understand objects as large as a planet, or as microscopic as a molecule.

Define Vocabulary Words

You can use clues in the text to help you understand a new word. You can also use word study skills. The chart below lists some of the most common root words in science. If you know these roots, you can unlock the meaning of many words.

Infer

Long before scientists landed rovers on Mars, they concluded that Mars once had liquid water. They made this inference because they had observed channels and flow features that were similar to water-formed structures on Earth. Inferences are interpretations about things you have not experienced—or cannot experience—directly. Inferences are based on knowledge or prior experience combined with evidence. If you see the garbage cans in front of your house knocked over and the trash spread around the street, you might infer that the neighbor's dog got into the trash. However, if there was a big windstorm the night before, you might make a different inference.

10 Science Root Words to Know

Root	Meaning	Examples
astro-	star	astronomy, astronaut
bio-	life	biologist, biome
cardio-	heart	cardiogram, cardiologist
geo-	Earth	geology, geography
hydro-	water	hydrosphere, hydroelectric
magni-	great, big	magnify, magnificent
micro-	small	microscopic, microorganism
-ology	the study of	ecology, biology
thermo-	heat	thermometer, thermal
zoo-	animals	zoo, zoology

Predict

A prediction is an inference about a future event based on current evidence or past experience. Scientists make predictions using their research and prior knowledge. For example, meteorologists make predictions about the weather. They analyze the current conditions and use their knowledge of weather patterns to predict what the weather will be and how it will change over a period of time. You make weather predictions as well. When you see large, puffy, dark clouds, you can predict that a thunderstorm is headed your way.

Analyze

Scientists analyze the data that they collect. This means they examine the data carefully and in detail. They look for patterns or trends that will help them make conclusions. Suppose you took a poll of the best-selling lunches at school during one week. You would have to analyze that information to reach a conclusion about what your classmates like to eat best. You could identify the bestsellers every day, and you could also see how many lunches were sold on each day of the week. Analyzing the information would enable you to draw conclusions about the most popular and least popular lunches.

Communicate

Communication is an essential part of science. When scientists communicate, they share ideas, information, results, and opinions with others. When you communicate, you talk, write, or listen. In the process, you share information and confirm ideas. You learn from other people when you communicate.

▲ You can make predictions when you look at the sky. If you see large, puffy, dark clouds, you know a thunderstorm is probably nearby.

THINKING LIKE A SCIENTIST

Summing UP

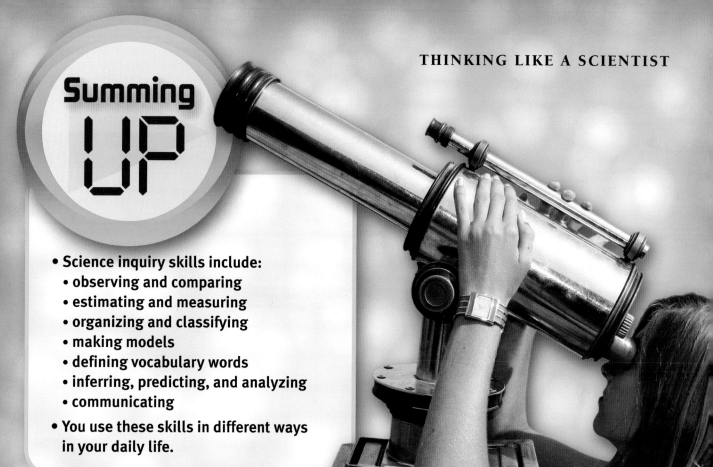

- Science inquiry skills include:
  - observing and comparing
  - estimating and measuring
  - organizing and classifying
  - making models
  - defining vocabulary words
  - inferring, predicting, and analyzing
  - communicating
- You use these skills in different ways in your daily life.

Putting it All Together

Choose one of the research activities below. Work independently, in pairs, or in a small group. Share your responses with the class.

1 Pick five of the skills described in this chapter and write an example of how you use each skill in your everyday life.

2 Work with a partner to think of an everyday problem that one or both of you recently solved. Describe the steps in the solution to the problem. Draw a flow chart of the solution process. Wherever appropriate, identify the science skills you used.

3 All of the skills in this chapter are important to the scientific process. Choose the one you think is the most important and write a persuasive paragraph to convince others of why this one stands apart from the rest.

Chapter 4
Science and Society

How does science help people make decisions about the world?

▲ The sea ice covering the Arctic Ocean has been melting earlier each year, making it harder for polar bears to hunt.

Scientific research helps society make decisions about public health, the environment, and industry. The work that scientists do can change people's points of view. Scientific research can lead to new laws and regulations. That has been true in the past regarding smoking, and it is true today regarding climate change.

Science and Public Policy

Not too long ago, scientific research caused significant changes in how people thought about smoking and the laws regulating smoking. In the 1940s and 1950s, medical researchers began to study how cigarette smoking affected people. At that time, almost half of the adults in the United States smoked.

In the early 1960s, the Surgeon General of the United States formed a group to examine all of the scientific research about smoking. The Surgeon General's job is to educate the public about health issues. After looking at more than 7,000 reports and studies from the medical community, the group determined that the evidence showed a link between cigarette smoking and an increased chance of lung cancer. In 1964, the Surgeon General issued a warning to the public about the dangers of smoking.

As a result of this report, the government required all cigarette packages to carry a warning about the dangers of smoking. The government also banned cigarette advertisements on television. The report led to campaigns to stop people from smoking.

Medical researchers continued to discover other dangers of smoking, such as heart disease and emphysema, a lung disease. Researchers also concluded that exposure to secondhand smoke was hazardous. Secondhand smoke can lead to cancer, asthma, and other diseases. State and local governments began to ban smoking in public places.

Scientific research spurred the government to pass laws regulating smoking. It also made the public aware of the danger of smoking. A study in 2008 indicated that the number of adults who smoked had decreased to below twenty percent.

▲ The Surgeon General's report led to warning labels on all cigarette packages.

▲ Smoking used to be allowed on airplanes, buses, and trains. Today, it is banned from all of those places.

CHAPTER 4

Meet the Scientists: Disease Detectives

The field epidemiologists who work for the Centers for Disease Control and Prevention are like disease detectives. When there is an outbreak of an unexplained illness, it is their job to figure out what is making people sick. These scientists travel around the world and spend weeks in the field collecting evidence. They identify new diseases, help doctors treat patients, and educate communities about stopping the spread of a disease.

Science and Change

Today, one of the biggest issues facing the world is climate change. The term *climate change* is being used more often than the term *global warming* because it conveys that other changes in addition to rising temperatures are occurring. Scientists are not entirely sure what causes climate change. Many believe that climate change is directly related to the increase in greenhouse gases in the atmosphere.

The main greenhouse gas is carbon dioxide. Since the early 1900s, the amount of carbon dioxide in the atmosphere has increased dramatically. This is a result of increased burning of fossil fuels—coal, oil, and natural gas—in factories and cars. When fossil fuels are burned, they release carbon dioxide.

Carbon dioxide is always present in the atmosphere. It helps to trap the sun's heat near Earth. Without it, Earth would be too cold for life to exist. But excess carbon dioxide can lead to a rise in global temperatures. While the detected increase of a few degrees over the last hundred years may seem small, continued increases in temperature could cause great changes on Earth.

The rise in temperature has already caused the ice at the North and South poles to melt at a faster rate. If this continues to happen, the melting ice could cause sea levels to rise, flooding cities along the coasts. The increase in temperature could also mean changes in global weather patterns. Scientists predict that some parts of the world will face severe droughts. Other parts may face more powerful and frequent storms.

Not everyone, however, believes that climate change is a result of human activity alone. Earth's climate has been changing for millions of years. Humans have been around for only the last twenty thousand years. And we have accurate climate readings for only the last one hundred years.

Science can present the evidence that supports climate change. But it is up to governments around the world and individual citizens to decide how to respond to the issue. They have to evaluate the scientific evidence and consider the best ways to address the problem. The more people understand science and the scientific process, the more able they are to participate in the debate around this issue.

▲ Scientists believe global warming caused the Wilkins Ice Shelf to break free from Antarctica in 2008.

SCIENCE AND SOCIETY

They Made a Difference
Mario J. Molina
(1943–)

As a young scientist, Mario Molina took on a big challenge. Chemicals called chlorofluorocarbons, or CFCs, were used in refrigerators and spray cans. Molina wanted to know what happened to these chemicals when they were released into the air. He found that CFCs drifted into the stratosphere. In this layer of the atmosphere, ozone gas protects Earth from the sun's ultraviolet rays. Molina discovered that CFCs destroyed ozone. He helped alert the world to the danger. In 1987, the nations of the world met and agreed to end CFC production. In 1995, Mario Molina was awarded the Nobel Prize in Chemistry. He is the first Mexican citizen to win a Nobel Prize in science.

CHAPTER 4

Citizens as Scientists

When scientists need help, they turn to the scientist in all of us. They ask volunteers to help gather data. For more than 100 years, thousands of birders have participated in the annual Audubon Society Christmas Bird Count. They fan out across the country to count and identify birds. Scientists use the data to study the long-term health of bird populations. Conservation groups use the results to create plans to protect birds and their habitats, or living areas.

Scientists are also asking citizens to gather data to help them study climate change. Project Bud Burst asks people to watch the trees and flowers in their area and to record when they start to bud. Frog Watch, which is sponsored by the Association of Zoos and Aquariums, asks people to collect data about the frogs and toads in their neighborhood.

Hands-On Science: Mini-Nature Watch

You do not need to travel to a beach or forest to observe the natural world. Try a fifteen-minute nature watch. Find a spot in your backyard or in a local park. You can even use the view from your bedroom window. Spend fifteen minutes just watching. What types of plants do you see? Are there any animals? At the end of fifteen minutes, write a description of the scene. Include as many details as you can. Accompany your description with an illustration.

Checkpoint: Read About It

Would you be interested in participating in a citizen science program? Look online to read more about the programs described here. Tell why you'd like to join one.

▲ Citizen science projects invite volunteers of all ages to participate in doing science.

SCIENCE AND SOCIETY

Summing UP

- Scientific findings play a role in decisions about public health, the environment, industry, and other issues in society.
- When science moves into the public arena, scientists must not only explain their findings, but also help the public understand the nature of science.
- The more people understand how science works, the more informed their decisions will be.

PUTTING IT ALL TOGETHER

Choose one of the research activities below. Work independently, in pairs, or in a small group. Share your responses with the class.

1 Scientists warn that climate change will greatly affect people's lives and public policy. Use books and Internet resources to find out how climate change may affect the area where you live. Then use your findings to describe how the decisions that public officials make locally and at the national level might be affected.

2 Work in a small group to choose a scientific discipline and create a time line of important developments in this field that have affected public policy in the twentieth and twenty-first centuries. Describe some of the developments. Explain how your time line illustrates that science is an ever-changing process.

3 For one week, collect all of the junk mail you receive. At the end of the week, sort and organize the mail. How many pieces of junk mail were received? Estimate how much junk mail you receive per year. What are some of the effects of this much wasted paper? How can you reduce this waste?

CONCLUSION
How Scientists

▼ No matter what they are studying, scientists use inquiry and investigation to find out more about the natural world.

WORK

Chimpanzees, volcanoes and particle colliders could not be more different. The one thing that they all have in common is that they are all subjects of scientific research. The scientists studying these various topics are from all over the world and have different backgrounds and interests. The one thing they all have in common is that they are dedicated to learning more about the natural world. Their research illustrates how science and scientists work.

The volcano scientists at the Alaska Volcano Observatory are trying to answer a question that was probably posed by the very first person to see a volcano: Why are smoke, ash, and hot rock spewing out of the top of a mountain? Although scientists today know more about why volcanoes erupt, there is still a great deal to be learned.

Through close observation of the chimpanzees, scientists in the Goualougo Triangle continue to make exciting discoveries about these animals. Much still remains to be uncovered.

How did the universe come into being? What is its structure and composition? The Large Hadron Collider may provide some answers. Scientists will test their ideas by conducting experiments in the particle collider and then analyzing the results.

The knowledge that these and other scientists have gained is a result of having creative, inquiring minds. Each one follows the scientific method and applies the fundamental science skills you've read about in this book. Whether they are working in a laboratory, traveling across the ocean, climbing through the jungle, traveling to the far reaches of space, or sitting next to you, all scientists contribute to our collective scientific knowledge. This knowledge helps to explain what happens in the world around us and can be used to benefit society.

How to Write a

Communicating the Results

Communicating the results of an experiment is an important part of the scientific method, and writing a lab report is one way to do it. A lab report allows others to repeat your tests, review your conclusions, and build on your results. It therefore must be as complete and accurate as possible. You will need to supply your preliminary observations, a description of the experiment you conducted, and the data you collected. Here are the steps to follow when writing a lab report.

Introduce your question or problem.

Describe the question or problem that your experiment will answer. Be sure to state it as a question. Explain why this question is important. What observations got you interested in the question? Mention any research that you have done on the topic. To write the report in the third person, avoid using words like "I" and "we."

State your hypothesis.

A hypothesis is a prediction that should be stated as an answer to the question. Based on all that you know about your topic, state what you expect to find.

List the materials you use.

Materials also include equipment. Be specific about the amounts whenever such information is important.

Indicate safety precautions.

Doing experiments is often exciting, but you should always conduct them safely. Be aware of any safety precautions you observe and indicate them.

Describe the procedure.

List all of the steps that you follow in the correct order.

Present your results.

You will want to present all of the data that you collect. Your data can be both quantitative and qualitative. Use tables, charts, and graphs to present the quantitative data. Describe qualitative observations.

Write your analysis and conclusion.

Use the data you collect to accept or reject your hypothesis. Explain the conclusion that you reach after completing the experiment. Restate your hypothesis and indicate if your experimental results support or fail to support your hypothesis. Think about the results of your experiment: What could you do differently next time? Is there a way to improve your experiment? Did your experiment generate any new questions?

Cite any references used.

You may have used books, magazines, newspapers, interviews, or the Internet to research your topic. Be sure to list the author, title, date, and place of publication or Web address for each source.

Sample Lab Report

Question or Problem
How does moisture affect the popping ability of popcorn kernels?

Hypothesis
If the amount of moisture in popcorn kernels is directly related to popping ability, then overdried popcorn will not pop as well as regular popcorn.

Materials
- 3 cups popcorn
- large bowl
- large baking pan
- oven mitts (*use when handling hot pans*)
- air popcorn popper
- measuring cup
- 3 paper towels

Procedure
1. Measure three level 1/2-cup batches of popcorn. Place each batch on a paper towel. Seal the bag of remaining popcorn. Label the towel samples A1, A2, and A3. Allow the popcorn to air-dry for 7 days. At the end of the drying period, record any differences in the appearance of the overdried and regular kernels. (You may also dry popcorn on a pan in a 93°C (200°F) oven for 90 minutes. Be sure to cool before handling.)
2. Design a table to record your data.
3. Pop the overdried corn in three batches. After each batch is popped, pour the popcorn onto a baking sheet and count the number of unpopped kernels. Record this information in the appropriate place in your data table.
4. Measure three level 1/2-cup batches from the sealed bag. These are samples B1, B2, and B3. Pop each batch of popcorn. Pour each batch of popped corn onto a baking sheet and count the number of unpopped kernels in the batch. Record this information in the appropriate place in your data table.
5. Add the number of Sample A unpopped kernels and divide by 3 to average the results. Record the average in your data table. Do the same for Sample B.

Data

The average number of unpopped kernels left after popping the dried corn was 15.

The average number of unpopped kernels left after popping the regular corn was 5.3.

Analysis and Conclusion
The experiment supports the hypothesis that overdried popcorn would not pop as well as regular popcorn. There were always more unpopped kernels in the batches of overdried popcorn. To further support the hypothesis, it would be useful to test more batches with exact kernel-counts. It would also be interesting to find out if the method of popping affected the number of unpopped kernels.

References
"Why Popcorn Pops and Other Grains Don't" Moments of Science @ indianapublicmedia.org

GLOSSARY

control group (kuhn-TROLE GROOP) *noun* a standard of comparison for checking the results of an experiment (page 21)

data (DAY-tuh) *noun* information from experiments and observations (page 21)

dependent variable (dih-PEN-dent VAIR-ee-uh-bul) *noun* what is measured in an experiment (page 21)

experimental group (ik-spair-ih-MEN-tul GROOP) *noun* where all variables in an experiment are tested (page 21)

hypothesis (hy-PAH-theh-sis) *noun* a statement that predicts the outcome of a scientific experiment or investigation (page 19)

independent variable (in-deh-PEN-dent VAIR-ee-uh-bul) *noun* the variable that is being tested in an experiment (page 21)

pseudoscience (soo-doh-SY-ens) *noun* things that appear to be scientific, but that do not meet the same standards as true science (page 13)

qualitative data (KWAH-lih-tay-tiv DAY-tuh) *noun* data that can be observed but not measured (page 22)

quantitative data (KWAN-tih-tay-tiv DAY-tuh) *noun* data that can be measured (page 21)

scientific law (sy-en-TIH-fik LAW) *noun* a statement that explains something that has been shown to be true through countless observations (page 12)

scientific method (sy-en-TIH-fik MEH-thud) *noun* the system for finding the answers to questions through observing and experimenting and testing ideas (page 16)

scientific theory (sy-en-TIH-fik THEER-ee) *noun* a body of well-tested observations that explains the natural world and predicts new findings (page 12)

spontaneous generation (spahn-TAY-nee-us jeh-nuh-RAY-shun) *noun* the theory that living things can come to life from nonliving things (page 8)

variable (VAIR-ee-uh-bul) *noun* an element that may change the outcome of an experiment (page 20)

Index

Alaska Volcano Observatory, 5, 43
branches of science, 14
Centers for Disease Control and Prevention, 38
citizen scientists, 40
climate change, 36, 38–40
control groups, 21, 23
data, 18, 21–23, 25, 27, 34, 40, 44–45
dependent variable, 21
Evans, Robert, 19
experimental groups, 21, 23
experiments, 8–10, 18–23, 25, 26–27, 31, 43–45
French Academy of Science, 9
germ theory, 12
Goualougo Triangle Ape Research Project, 4, 43
hypothesis, 19–20, 22, 24–25
independent variable, 21
lab reports, 44–45
Large Hadron Collider, 4, 43
Laws of Motion, 13
Leeuwenhoek, Anton van, 9
microorganisms, 9–10
models, 32, 35
Molina, Mario J., 39
Mount Redoubt, 5
Newton, Isaac, 13
Pasteur, Louis, 10–12, 16
pseudoscience 13
qualitative data, 22–23, 44
quantitative data, 21
Redi, Francescwo, 8–9, 16
science root words, 33
scientific law, 12–13, 15
scientific method, 5, 16, 18, 23, 25, 28, 43, 44
scientific theory, 12–13, 15
smoking, 36–37
spontaneous generation, 7–10, 12
Surgeon General, 37
variables, 20–21